Help Your Child

Furry Foxes

RICHARD & NICKY HALES
and ANDRÉ AMSTUTZ

GRANADA

Published by Granada Publishing in 1984
Reprinted 1985
Granada Publishing Limited
8 Grafton Street, London W1X 3LA

British Library Cataloguing in Publication Data

Hales, Richard
    Furry foxes. – (Help your child to count; 4).
    – (A Dragon book)
    1. Numeration – Juvenile literature
    I. Title   II. Hales, Nicky   III. Series
    513'.5   QA141.3

        ISBN 0 246 12465 2 (hardback)
        ISBN 0 583 30726 4 (paperback)
        Printed in Spain by Graficas Reunidas

Help Your Child To Start Maths

Once mathematics was the most difficult subject for children to learn. Parents would not dare do more than teach their toddlers to count to ten. Sums were for school. But maths need not be daunting; it is a part of our everyday lives. We use it in travelling, shopping and cooking. Children first meet maths in the home: in songs and rhymes, through helping in the kitchen, by playing games and just by talking about the things around them.

With your help and encouragement your children can absorb mathematical concepts while simply having fun. This series of books will help you talk about maths and suggest activities and games that will make maths child's play for you and your child.

Here are five foxes

They are called
the furry five

1 and **1** and **1** and **1** and **1** make **5**

The furry five love fun and games

Who tore the carpet?
Who chewed the socks?
Who bit the slippers?
It's a furry fox!

Who is behind the sofa?
Hiding in a box?
Jumping on the table?
It's a furry fox!

Three furry foxes
hiding from their mum.
Two furry foxes
filling up their tums –

three and two make five.
3 and **2** make **5**

Two furry foxes
with boots upon their feet,
two furry foxes
cycling up the street,
one furry fox
learning how to drive –

two and two and one make five.

2 and **2** and **1** make **5**

The furry five
have fun at school.
Four are painting pictures.

One is playing the fool!
Four and one make five.

4 and **1** make **5**

Five furry foxes
fighting on the floor,
and if one furry fox
should roll right out the door –
there'll be
four furry foxes
fighting on the floor.

Three furry foxes
playing in the bath,
one furry fox
having a good laugh,
one furry fox
about to take a dive –

three and one and one make five.

3 and **1** and **1** make **5**

Four sleepy heads
comfy in their beds,

one sneak on his feet,
pinching more to eat –

four and one make five.
4 and **1** make **5**

What's this, a midnight feast?

How many foxes are in bed?
How many foxes are upside down?
How many foxes are eating?

Did you hear what daddy said?
Furry foxes back to bed!

Things To Do

Colour I Spy
Use colours rather than letters to play I Spy, saying 'I spy with my little eye something that is red.'

Matching
In this game you give your child, say, five dolls or soft toys and five sweets. Then you ask if there are enough sweets for the dolls to have one each. The child finds out by matching one sweet to one doll.

You can change the number of dolls and sweets, and use toy figures, model cars, counters and egg cups as you choose.

Sorting
Ask your child to sort various buttons, counters, toys or bricks into groups of a single type. For example, ask him to make a set of all the buttons with two holes in them or all the red bricks, or all the toy animals.